# HEALTHY
## FOOD CHOICES

Diana Noonan

# CONTENTS

Food Is Fuel ........................ 3

The Secret's Inside ............... 5

Balanced Choices ................ 8

Food Chains ...................... 12

Food Crops ....................... 14

Meat ................................. 17

Try Something New ........... 22

Making Choices ................. 23

Glossary ........................... 24

# FOOD IS FUEL

Everyone likes to eat some foods more than others. But did you know that the choices you make about food are decisions that will affect how your body will grow?

Food gives you the fuel that you need to grow. Some foods have more **nutrients** in them than others. Making good choices about what you eat will help your body, including your bones and your brain, to grow well. The food you choose could even make you smarter!

Did you know that growing brains need certain nutrients from food? "Brain foods" include fish, vegetables such as broccoli and cauliflower, egg yolks, and milk. Other smart choices for your brain include nuts and meat.

A particular kind of food may taste good, but that doesn't mean that eating it is the best choice for you to make. Sweet, sugary foods like cookies or a piece of cake taste good. Eating these types of foods once in a while is not a problem, but if you eat foods that are high in sugar and fat all the time, it is bad for you. This is because your body uses everything you eat as fuel, and these types of foods are filled with too much of the wrong kinds of fuel.

> How much does the average person in the United States eat in his or her lifetime? No one is quite sure, but estimates vary between 1 ton (about 22,680 kilograms) and 70 tons (about 1,587,600 kilograms)!

# THE SECRET'S INSIDE

After you eat something, your body breaks food apart so that it can use what it needs. Some foods have more healthy nutrients in them than others.

## Foods High in Nutrients

Starch comes from plants. Your body uses starch for energy.

Protein comes from plants and animals. Your body uses protein to grow and repair itself.

**Vitamins** and **minerals** come from many different foods. They increase your chances of not getting sick.

Sugar comes from plants. Your body uses sugar as a source of quick energy, but it can't cope with too much sugar. Excess sugar is stored as fat and contributes to some serious health problems.

Sugar cane

Fat comes from both plants and animals. Your body uses fat for energy and warmth, but some kinds of fat are healthier to eat than others.

Fish contain fats we need.

The fact that you like to eat a **variety** of foods is your body's way of telling you that it needs many different nutrients—not just one or two.

Did you know that eating tuna or peanut butter sandwiches can help you to get better test scores? Your brain needs the protein in these kinds of foods.

# BALANCED CHOICES

A balanced diet means eating plenty of different kinds of natural foods every day—with all their different nutrients. Here's a fun way to achieve this—eat lots of different colors!

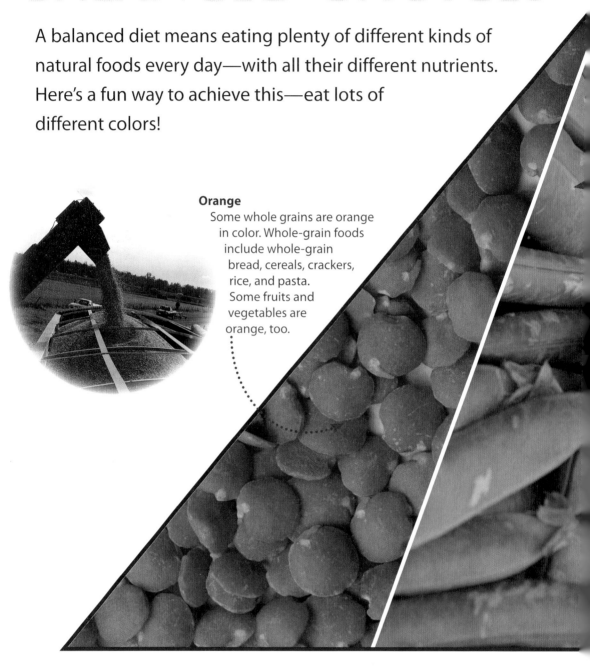

**Orange**
Some whole grains are orange in color. Whole-grain foods include whole-grain bread, cereals, crackers, rice, and pasta. Some fruits and vegetables are orange, too.

**Green**
Dark green leafy vegetables, avocados, beans, and peas are all filled with healthy nutrients.

**Red**
Eat a variety of fruit and vegetables with red skins. Meat is often red, too.

**Yellow**
Oils come from grains, nuts, olives, and fish. Vegetable oils are often pale yellow.

**White**
Calcium-rich foods, including dairy products, are white—like milk. Fish and poultry are sometimes white, too.

**Purple**
Some beans, fruits, and vegetables are purple!

What other colors can you find in your food?

# Super Foods

Some foods are so healthy to eat that **nutritionists** call them "super foods."

**Grains and nuts**
Beans
Soybeans
Oats
Walnuts

**Dark-colored fruits and vegetables**
Apricots
Avocados
Broccoli
Blueberries
Pumpkin
Tomatoes

**Lean meat**
Skinless turkey breast
Tuna
Sardines

# FOOD CHAINS

Where does our food come from? Most plants make food by changing sunlight into food. Plants are called producers because they produce food from the sun's energy. The sugar they make is the fuel they need to grow.

Consumers eat plants and other living things that eat plants.

When you eat a hamburger, you're consuming meat that comes from steers. Those steers ate plants for their fuel, and then their meat is fuel for your body. The food you consume gives you energy.

Decomposers eat plants and animals after the plants and animals have died. **Fungi**, **bacteria**, and animals such as earthworms are decomposers. They release nutrients from dead plants and animals back into the soil, where they are used once again by growing plants.

The movement of energy from producers to consumers to decomposers is called a food chain. Food chains link together to form a food web.

**A food chain**

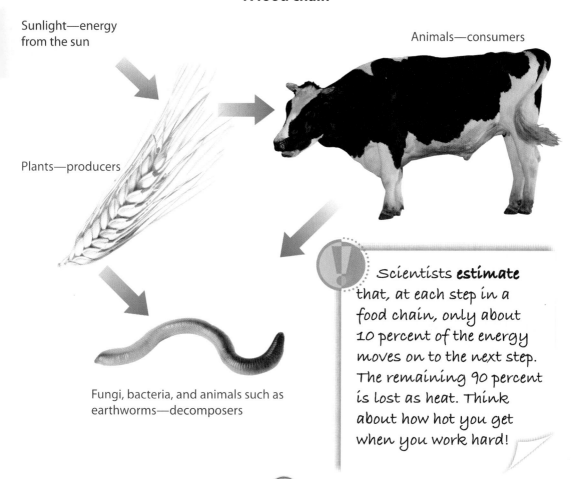

Sunlight—energy from the sun

Animals—consumers

Plants—producers

Fungi, bacteria, and animals such as earthworms—decomposers

Scientists **estimate** that, at each step in a food chain, only about 10 percent of the energy moves on to the next step. The remaining 90 percent is lost as heat. Think about how hot you get when you work hard!

# FOOD CROPS

Plants that are rich in energy provide the **basis** for much of the food we eat. Energy-rich crops include grains like wheat, corn, and rice. When we don't eat these plants ourselves, we sometimes use them to feed the animals that we eat.

Root crops like potatoes and yams are also rich in energy. Root crops are parts of plants that grow underground that you can eat. Examples of root crops are turnips, beets, sweet potatoes, and carrots. They are called "root crops" because they form part of a plant's root system.

Nutritionists recommend that these energy-rich plants make up the largest part of your diet.

## Sprout an Energy-Rich Food Plant

Here's how to see how one energy-rich food plant starts growing.

### You will need:

- an avocado—any kind will do, but the small, dark, bumpy-skinned Florida avocados **sprout** the fastest
- a plant pot
- potting mix—soil that is rich in nutrients
- water
- a clear plastic cup
- a sunny windowsill

### What to do:

1. Eat the avocado and save the seed.
2. Wash off the seed and leave it overnight to dry.
3. Fill the pot with soil almost to the top.
4. Push the seed into the soil, big end down, until it is two-thirds covered by soil.

5. Water the soil in the pot.
6. Turn the plastic cup upside down and put it over the seed—this will help the seed to stay moist and warm.
7. Put your pot on a sunny windowsill.
8. Keep the soil moist and keep a record of what happens.

| Date | Observation |
|---|---|
| 11/4 | Still just an avocado seed! |
|  | The seed has split! |
|  |  |
|  |  |
|  |  |
|  |  |

9. If your avocado plant gets too big for the pot, you will need to put it in a larger pot, or you could plant it outside.

**Compare your results with a friend:**
Ask a friend to do this experiment at the same time. Compare your observations. Do you get the same results?

# MEAT

In a food chain, it takes about 8 pounds (3.63 kilograms) of corn to make 1 pound (0.45 kilograms) of meat.

However, if the animal did not eat that corn, think about how many people 8 pounds (3.6 kilograms) of corn could feed!

Tortillas are made from cornmeal.

Here's how much meat farmers can produce from 1 bushel of corn.

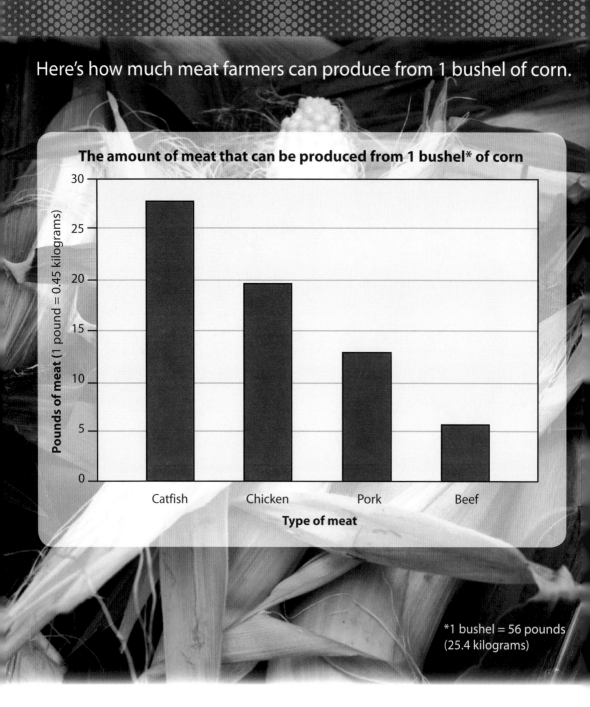

As you can see, it is more **efficient** to produce corn-fed catfish than corn-fed beef. However, it is even more efficient to eat the corn yourself!

# Lean Meat

Meat is a good source of protein. Lean meat is meat with little fat on it. Lean meats include the meat from fish and from poultry, such as chicken and turkey.

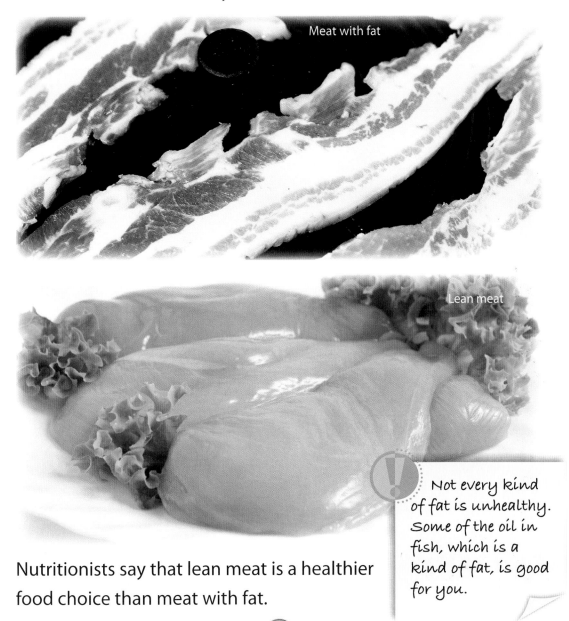

Meat with fat

Lean meat

Nutritionists say that lean meat is a healthier food choice than meat with fat.

Not every kind of fat is unhealthy. Some of the oil in fish, which is a kind of fat, is good for you.

There are different ways to raise lean meat. For example, the way chickens are raised affects how lean or fatty they are. Chickens that are raised in cages are fed grain. Since they are unable to move around freely and don't get much exercise, their meat is fattier than the meat of chickens that are raised in other ways.

> Did you know that some fish are also raised in cages? They are fed food pellets.

Free-range chickens don't live in cages and have more room to move around. They have enough room to behave naturally, scratching the ground for insects and seeds to eat. Because free-range chickens get more exercise, their meat is leaner. Since the food free-range chickens eat is more varied, their meat contains a wider range of nutrients.

# TRY SOMETHING NEW

Growers sometimes sell their fruit, vegetables, and other food at farmers' markets. Some growers even have roadside stands where you can buy fruit and vegetables fresh from the field. Some farms, farmers' markets, and stores sell interesting foods you might not have tried before. There may be more ways to eat a wide range of nutrients than you realize!

Pike Place Market, Seattle

# MAKING CHOICES

Once you know the nutritional value of different kinds of food, you can make smarter choices about what to eat. You'll be in for a surprise—healthy food tastes good, too!

"Eat your vegetables. They're good for you!" Has anyone ever said this to you? Well, it turns out they were right!

Do you know which vegetable is the world's most nutritious? Some nutritionists say it is broccoli! Number two may be spinach. Broccoli contains important nutrients, including vitamin A and vitamin C, and spinach is a good source of iron.

# GLOSSARY

**bacteria**—tiny, single-celled organisms

**basis**—something upon which another thing is based

**efficient**—with less waste of energy

**estimate**—calculate

**fungi**—organisms such as mushrooms, toadstools, moulds, and yeasts

**minerals**—salts that plants get from soil and that your body needs for good health

**nutrients**—the substances in food that help living things stay alive and grow

**nutritionists**—people who study what we need to eat

**sprout**—grow leaves

**variety**—a range or mixture

**vitamins**—substances created by living things that your body needs in tiny amounts for good health